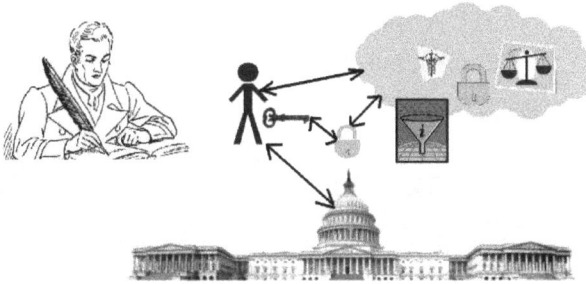

"The User Guide to

Gov 2.0 - Federal"

By

Stephen Imholt

DEDICATION

This book is dedicated to three people, Bob Woodson, Paul Ryan and Art Burton.

This may seem an unlikely combination of people but there is a commonality, at least to me.

Art Burton is a community activist. He says he's left that kind of thing behind, but if you emphasized the word community, and then said "Who will help to make a community?" Art is the guy. This is currently true for the Mosby areas in East Richmond. The work he is doing at kinfolks deserves its own book. I hope someone with real talent takes the time to write it.

This pamphlet is also dedicated to Bob Woodson, who recognized that change has to be personal to be successful. Change has to respect those we are expecting to change, and that change must be one person at a time. Those changes are the most lasting kind of change. His Center for Neighborhood Enterprise is based on that concept. The concept of having the community "own" the Federal Neighborhood Office was based in part on those same concepts.

Lastly, this pamphlet is dedicated to Paul Ryan. Congressman Ryan unafraid to reach out and incorporate some of Bob's principles into his grant proposals structure. It takes guts to actually talk about changing government, even at the local level, when large numbers of conservatives simply want smaller government. I'd like these books to follow their lead.

FORWARD

If you think the government we get from Washington, DC is the government we deserve, then you should stop reading here, because this entire series of books is likely going to aggravate you.

A lot.

But, if you are actually open to having things work better, you might want to read this book.

Think of this as a User Guide for an improved way of having government work for you, even though the improvements haven't even been built yet.

This user Guide is for anyone who ever built a Barbie Doll #@$*&#)(Dream House[1]. It is for people that need to know how to get more from the government as far as quality, responsibility, and effectiveness through a redesigned federal delivery system. But most especially it's for people that don't have a law, political science or information technology degree.

It's for Joe Everyman, the people who actually built this country.

While this may seem like a guide to a fantasy, it is very, very doable. But only if we have the will to change.

---

[1] The Barbie "#@$*&#)(Dream House is a product of the Mattel Corporation. The version sold thirty years ago should have been classified as an instrument of torture in the toy assembly category. The Federal Government has the same kind of torture. It is called the Federal Tax Code.

TABLE OF CONTENTS

Federal Government 2.0 The User Guide
Introduction

ABOUT THIS BOOK AND GOVERNMENT 2.0

CRITICAL NOTE: The book is written as if the first pieces of Gov 2.0 are running and the remainder are being brought on line just like new versions of the iPad. That is, the user guide is written as though a new release will occur every couple of months. That's not really the case, but more of the hope.

This book is part of the series of books for Government 2.0 or Gov 2.0 for short.

This book is meant to be like the book that you get when you buy a new car. Or perhaps that getting started guide that comes at the beginning of the user guide for a computer is another good description.

This user guide will tell you how to do something using Gov 2.0.

Gov 2.0 is a term used to describe an entire collection of changes to government. These changes allow us to improve the effectiveness in delivery of government services at a reduced cost.

Gov 2.0 is not just about improving services, and reducing costs.

Gov 2.0 is about how to use technology to protect our rights.

Gov 2.0 is about how to improve how the government works.

---

[2] This book assumes that the vision and architecture contained in Volume design and implementation plans contained in Volume 2 of Government 2.0 either have been adopted or are in process of implementation.

Gov 2.0 is about how to make these changes in a way that lets ordinary people control their government at the local level.

This book doesn't explain how everything is interwoven. How things are interwoven is contained in the other books which comprise the Government 2.0 Series especially the first two books, Gov 2.0 Vision and Architecture and Gov 2.0 Design and Implementation Planning[2]

If you only want to know what is Government 2.0 going to do for you, or just as importantly to you, then this book is for you.

If on the other hand you are more interested in knowing the why and the how of Government 2.0, then read this book, then proceed to read volume 1 and 2 of the series.

Each of those volumes has identified at critical spots where the average reader can skip the more in depth pieces of the new architecture. The detail pieces can at some levels become rather complex, and frankly dry as a desert. I hope that splitting up the books that way makes it easier not more difficult to understand the concepts.

This book like the others in the series actually invites criticism. So if you come across an area in the books which is unclear or which you think can be improved upon, please let me know.

If you want to raise an item as you read or after you finish the book, the final section describes some ways to provide feedback. The only thing that is asked of you, if you want to criticize the contents, at least read the contents. Is that too much to ask?

INTRODUCTION

The good news is that the US government is the first government that was actually intentionally designed instead of developing organically.

The bad news is that the design is now over 225 years old. While the concepts are sound, the delivery and management functions of Federal Government are badly in need of updating.

In 1791 people did not have the charge card, the cell phone, or the internet. In 1791 not all people had equal rights. Some had no rights at all. Some components of what we call individual liberty today only exist as a result of the Bill of Rights. When those 10 items were originally written and passed they only applied to today what we would call Old White Guys. Seeing those rights as belonging to other groups such as slaves and women came later.

The major problem with the government of 1791 was that it distinguished between different categories of people. Indians didn't count at all. Slaves only counted for three fifths of a person. Most personal rights were only sprinkled through the constitution or more likely ignored altogether. It didn't found a government with large delivery systems at all. Those delivery systems grew over time. The Constitution wasn't set up assuming communications at the speed of light.

This book is about the way that Gov 2.0 will bring services to you in a more effective way.

It's about how Gov 2.0 uses technology to improve on the delivery of government services.

It's about how the technology you use every day has, and can continue to transform government.

You can read how Gov 2.0 assists neighborhoods in using targeted programs to meet their needs.

You can look up how people use Gov 2.0 to protect their privacy.

And how Gov 2.0 actively protects and enforces your rights.

Lastly, this book is about fostering neighborhood responsibility. But to understand that aspect of the book, you'll need to read a lot more of the Government 2.0 series.

SOMETIMES A NEW PERSPECTIVE HELPS

Because our country is based upon rights and freedoms, it makes sense to want to know what effects technology can have on those freedoms. The risks our rights have **from** technology are obvious. We are seeing them in headlines and on talk shows. Politicians campaign on them. We try to protect our rights from identity theft, hackers, secure networks, password theft, data protection etc. But, if the core rights and freedoms are to survive, we must start to consider what technology can do to **protect** our rights as well.

The ability for each of us to have a voice in our government is as fundamental to our rights, as is freedom of speech and of belief. The traditional view of a New England town meeting was a great example of how that freedom was exercised. Unfortunately, that isn't true anymore for the US Congress or the state houses in our larger states. It's not even true for City Hall in our larger cities. Our large population has clearly reduced the ability to interact and direct our own government. We are losing those rights if for no other reason because of the sheer numbers of citizens. Government has lost touch with people. Most people in fact don't even think they can have any effect on government.

Two components of Gov 2.0 are ways to use technology to begin to regain that right at the local level. One is the Federal Neighborhood Office (FNO). The other is the Federal Local Magistrates Court (FLMC). These components use technology to protect our rights. Using technology to preserve our rights is a core concept of Gov 2.0

These tools along with the Personal Data Store and the Federal ID Card help support reforming the Postal Service, the IRS and other government agencies. All of these items are part of the entire Gov 2.0 series of books.

HOW THIS BOOK IS ORGANIZED

This User Guide addresses things by topics. These topics are

- Why a User guide for a system which doesn't exist
- Installing Gov 2.0 in your area
- Using your New FNO
- Using your new Personal Data Store
- Managing your new Federal Neighborhood Office
- Using your new Federal Neighborhood Magistrate Court
- Other Everyday functions in Government 2.0
- Help Functions in Government 2.0
- Receiving Future Releases

Remember, Gov 2.0 has not been implemented. As of the fall of 2014, it is still in the virtual vision world of the developers and architects.

WHY A USER GUIDE WHEN THE SYSTEM DOESN'T EXIST

Let's be blunt. If and when we actually start to make Gov 2.0 actually become real all fifty states, this manual has no purpose. So in one respect, it would seem that you've just invested $4.99 or $2.99 or .99 cents for the electronic version, or even crazier, $6.99 for the paperback version for nothing. But don't feel bad. Part of your payment for the book is going to provide funding to further the goals and objectives of Government 2.0.

Unlike Jerry Seinfeld who had a comedy series about nothing, my book is about something. The fact that it doesn't yet exist, is really just a timing problem.

So why have a User Guide? Actually the User Guide helps some people to understand what all the ideas and concepts in the other Gov 2.0 books will look like, once actually implemented them. And given the history of the last 200 years do you really think people would even consider something like Government 2.0 if they didn't know what it would look like? Of course not.

Even with the User Guide, the largest hurdle is convincing people believe it's even possible. The effort to convince people it is even possible may take more than just a single political campaign.

The effort just to convince people it's even possible will take no less than a year or more likely two. And after you convince people it's even possible, it will take huge efforts by many people to make it go from vision to even a hope. It will take even more people to take the ideas from a hope, to a commitment. It will take even more people to go from a commitment to a movement.

But don't worry. This book is short enough that you can read it within an hour or two. I hope that where you bought it there is a return policy so you can get your money back. But even if you can't get your money back, it's worth it. Ask yourself "How much has the federal government told you that isn't real either?" and then "Does the government offer refunds"

When you stop laughing, if you still want your money back, just follow the return instructions on the web site (once I get enough money put together for there to even BE a web site). If there isn't yet a web site, then you likely had to buy it through Amazon or whoever else I get to publish this, and just follow their normal return procedures.

If you aren't able to read this book within an hour or two, please accept my apologies for letting our public education system fail you.

If you think I am getting rich writing this book, please think again. I believe it will take a bit over 6,000 e-book versions at 99¢ to get my 6,000 estimated hours of effort back at a dollar an hour. My e-book versions would have to sell over 45,000 copies for me to make minimum wage. While that would be wonderful, I don't expect that this will happen. After all, this book wasn't written to be the next Hunger Games or Harry Potter book.

Yet, there is not a single thing in this book that doesn't actually have real possibilities for people to make their government function effectively for them. If that is true, it is well worth the price.

If everything described in these series of books were built, it would be the first government program to cost less than originally estimated. That's not true of Social Security or Medicare. It's certainly not true of the Affordable HealthCare Act. I believe no other government agency founded in the last hundred years has cost less than expected. Gov 2.0 may even reduce the total spending by the federal government.

Even if people are not convinced that Gov 2.0 is THE answer, hopefully, it will convince them to start thinking about how to make the government easier to use.

The Guide can also help the general public understand how we can make government work better. The Guide could cause people to demand better from the government, and perhaps that will lead to candidates, and ultimately elected officials who understand the need to evolve our delivery systems.

INSTALLING GOVERNMENT 2.0 IN YOUR AREA

For people living in urban and suburban areas, the first inkling of how Gov 2.0 works is likely going to be when the Federal Neighborhood Office opens near you. So what is an FNO?

FEDERAL NEIGHBORHOOD OFFICE (FNO) -

The Federal Neighborhood Office (FNO) is a new kind of office of the federal government. First of all, it actually is built and delivered to the local neighborhood. Overall, its size is likely to be only as big as a small apartment building. That's why most people won't believe it can offer the services being described, as contractors begin the installation. It is then that you will know that implementation of Gov 2.0 has begun.

The FNOs are small enough to be more personal. Most people won't have to leave their neighborhood to get to an FNO in most cases. Once the FNO opens for operations, you will probably get to know the people working at your FNO.

The FNOs effectively is your local help desk for ALL federal services. It is just one part of the effort to streamline, and modernize the public facing part of the Federal Government. It will support the Customer Service functions of the Social Security Administration, Postal Service, IRS, and most other federal offices.

The FNO will be a one stop shopping for the local community to access federal services

14   copyright Steve Imholt

The FNO will normally be run by one or more Government Authorized Agents (GAAs). These GAAs are contracted companies, sometimes even small companies from the local area. The GAAs report to a locally elected board. That's called the FNO Board. More about GAAs and the FNO Board are found in Managing Your New Federal Neighborhood Office.

NOTE: FNOs will open on different dates over several years

Since there are so many FNOs to open, not all FNO will open on the same date. Given that, the Federal Information Service will provide a web site so you can see when you are scheduled, and an entry page where you can tell the FIS when you want to be notified that the FNO Project for your area has begun.

Once operations at the FNO are actually running normally, the local community will have elections to choose their own board to manage the FNO. (See Managing You FNO section later in this book)

By then the community will have had a chance to get introduced to the staff (probably less than 10 people altogether) that are the local face of the federal government.

USING YOUR NEW FEDERAL NEIGHBORHOOD OFFICE

Your local federal Neighborhood office has a wide variety of support functions. These include:

- **US Postal Service email support**
- **US Passport Issuance from the Department of State**
- **Social Security Administration assistance**
- **Department of Veterans Affairs Assistance**
- **Internal Revenue Service Assistance**
- *Federal Neighborhood Office Registration and Support which includes*
  - *Federal Personal Data Store setup and maintenance support*
  - *Federal ID card Issuance and Maintenance*

The first three services will be available for every FNO opening. The remaining items may not yet be available when your FNO opens. The sequence of the agency projects, will find when those services start.

As agency projects are completed, functions are moved from the central offices in DC to the FNOs. As that happens, the user guide will also be updated.

The items listed in italics in the list don't exist within the federal government today. The items listed in bold do have something like it in the existing agency today, but they are being replaced by the new delivery service. The transitioned functions are described first.

USPS FUNCTIONS PERFORMED BY THE FNO

The Post Office processes available in most post offices today are being merged into the FNO. These common processes are transferred to the FNO as each FNO opens.

MAIL

Just like todays Post Office you can drop off mail at the local FNO. But you also have other options. [3] For those areas, Scan and Send (SNS) or Priority Mail is necessary to insure next day delivery.

SCAN AND SEND (SNS) EMAIL

For addresses listed in the USPS Email Directory, you can use Scan and Send. Scan and Send is a self-service function At the FNO you can scan and email it to their USPS email address. As FNOs are added the size of the USPS Email Directory will grow until eventually it will cover all US residents. [4]

SERVICED SCAN AND SEND (SSNS) MAIL

---

[3] Be aware, that if the person you are sending physical mail is living in an FNO area, it's likely not going to get there as fast as today's mail does.

[4] For those with home computing capability, you can of course email to those individuals documents directly to their USPS Email Address. This service is provided without charge to the sender or the recipient.

Similar to SNS Email, the sender can submit the letter to be emailed to the recipient. A GAA employee at the local FNO will then do the scanning and sending. This fee will be at rates determined by the local FNO Board but not more than $1.00 for the first page and .15 for each added page.[5]

PHYSICAL EMAIL DELIVERY

For people who do not have computers the FNO will print their emails. New USPS Emails will be printed and directed to the individual's physical address. The service is free for the disabled or for those with religious objections. Otherwise, the rate will be reasonable. It will not be greater than the first class mail rate.

PHYSICAL MAIL DELIVERY

When the last residents currently serviced by a local postal office are now within an operating FNO, mail service will be moved to the selected GAA vendor.

US PASSPORT SERVICES THROUGH THE FNO

---

[5] Rates may well be set at competitive rates for the local area. Leaving this as one of the local board controlled prices allows for communities which don't have a wide exposure to technologies not to be unfairly penalized

Once you are in an FNO area, you can apply for a new passport at your FNO. Unlike today's service you can do most of the application electronically. Using one of the public service access points (set up in small cubicles, just like in high school) you can complete the application form on line real time.

One of the staff members of the FNO will certify that your ID and other information actually is you. From there, the entire application is filed. Unless you are applying for a diplomatic passport, you should receive your passport in about 4 weeks.

Special additional processes have been dropped, so they won't extend wait periods.

Under the old system for passports, the time was a minimum of 4 to 5 weeks.

SOCIAL SECURITY ADMINISTRATION FUNCTIONS SUPPORTED BY THE FNO

While researching the Gov 2.0 books, I found something interesting. Most SSA public can be performed on line. The big exceptions to this are hearings and disputes[6]. If a person can't go online they have to go to the nearest SSA office. Local SSA offices focus on dealing with issues and disputes on the specific Social Security services for which the individual is eligible.

---

[6] A significant portion of these disputes will become part of the jurisdiction of the Federal Local Magistrate Court.

copyright Steve Imholt

With Gov 2.0, those who can't go online from home can use the kiosk at the local FNO. The local FNO will support most of the SSA operations previously performed by local SSA offices. These include operations like replacing your card.

Disputes with the SSA have been transferred to the Federal Local Magistrate Court described later in this guide.

DEPARTMENT OF VETERANS AFFAIRS ASSISTANCE

Today all VA Services are offered through separate VA facilities. These locations are spread throughout the country. Yet still some veterans have to travel long distances to receive these services.

Those portions of the job of the VA specific to their care will continue to be delivered in separate VA facilities. Some services such as VA Loans and Educational Assistance are being moved to the local FNO. This allows veterans to integrate into their community much more easily. It also allows veterans more freedom in deciding where to live. This could be useful when considering their next career move.

VA medical records will be moved to the vet's PDS. You can read more about the PDS in the next section of this book.

As the Veterans Affairs administration is decentralized, more of those functions will be transitioned to the Federal Neighborhood Offices.

IRS SERVICES THROUGH THE FNO

IRS Services through the FNO are described in the section Using Your New Personal Data Store.

WHEN AND HOW ARE FEDERAL NEIGHBORHOOD OFFICES OPENED

The chart below will show the calendar of major events in the creation of each FNO. The chart makes a lot of assumptions but the single biggest effect on the timeline to open up each FNO is the time to acquire and fit out the location. With over 70,000 locations, a new federal agency was created called the Federal Information Service (FIS). The FIS will have developed standard blueprints for pre-manufactured office buildings to minimize construction time and cost. For situations with extreme conditions or where real estate prices prohibit such a large footprint, the General Service Office (GSO) will acquire existing real estate for GAA vendors to fit out the FNO location.

copyright Steve Imholt

Day	Activity	Responsibility (s)
1	Request for Bid issued	FIS
30	GAA Bids Received	GAA Vendors
60	FIS Selects GAA contractor	FIS
75	FIS GAA Contract signed	FIS, GAA
120	FNO Building Manufacturing done	GAA
180	Fit out Completed	GAA
	Initial Notifications Mailed to local residents	GAA
210	Staffed Trained- Soft Launch	GAA
	Updates Mailed to local residents	FIS
240	Official Opening	FIS

The initial notices and updates will contain information about the initial set of the functions that the local FNO supports within the physical local office and a brief description of the major things that the FNO itself can better assist the residents of the area. In fact, these notices may well include a real version of this User Guide.

FNO REGISTRATION AND SUPPORT

In today's government (Gov 1.0) you may often feel as though you are treated like a number. While it may look like Gov 2.0 will only make it worse, the FNO is designed to actually personalize the service you receive from the office itself. Normally registering at the FNO is a once in a lifetime activity, and consists of the following steps, which for most people will occur shortly after the FNO opens for operations.

When you arrive at the FNO you will be greeted by the concierge. Once the concierge knows that you haven't been set up yet, he'll ask you how you want to get started. Basically, he'll ask if you want to set yourself up or do you want to have help from some of the FNO staff who function as Service Reps. During the initial setup, some Service Reps may be remote, but can help new members at the kiosk.

In either event, the following steps complete the initial setup and registration.

    a. A brief interactive video presentation of the FNO.[7]
    b. Validation of the residence of the individual within the FNO

  c. Issuance of the Federal Identification Card
  d. Creation of the Individuals Personal Data Store
  e. Video Presentation on the new US Postal Delivery
    mechanisms[3]
  f. Video Presentation on Using your Personal data Store[3]
  g. Setup of your Personal Data Store
  h. Setup of your PDS Email Account
  i. Setup of your Medical Records Wing
  j. Setup of your Veterans Administration wing[8]
  k. Setup of your IRS wing. [4]

Actually, once you created your personal PDS you could do the remainder of these tasks from home.

---

[7] These presentations are also available through both YouTube and FIS Websites and as links from within your PDS.

[8] Once the agency processes have been modified for Government 2.0PDS

       copyright Steve Imholt

Validation of Residence within the Federal Neighborhood Office

In order to receive services from the FNO you must either be registered at the FNO, or have already received a Federal ID Card.

Validating your Residence can be done either by providing a valid form of ID or by personal reference. These two options are explained as follows.

Providing a valid Form of ID

You can use an existing form of ID which includes:

Driver's License
State Issued ID card
Voter Registration Card
Utility Bill Showing your Address
Medicaid Card
Or any other Government identification showing your address

Or

Personal Reference

You may also receive a Federal ID Card if your identity and residence are confirmed by two members of the local FNO confirming your name and place of residence. The identity card will also identify whether you are a validated US Citizen. [9]

---

[9] You cannot access to the local voting functions unless you are a validated US citizen

Normally this would be used in the case of parents with children who are going to receive their own Federal ID Card and Personal Data Store.

CREATION OF YOUR NEW INDIVIDUAL DATA STORE

When you first receive your Federal ID Card, the representative will ask if you have a preferred PDS vendor or if you want to receive the default PDS vendor for the local FNO.

Regardless of which PDS vendor you select, the process to initialize and establish the link between your Federal ID Card and your Personal Data Store is the same.

Using the screen established, the service rep will actually create a new Personal Data Store for you, and will tie your Federal ID Card to the PDS. In addition, the service rep will give you a onetime pass code to be used at your convenience in managing your PDS.

Once the PDS is created, linked with your Federal ID Card, and the one time pass code has been issued, you now officially have an Individual PDS. More information about how to access your PDS is in the next section. It has information about how to use your one time pass code.

USING YOUR NEW PERSONAL DATA STORE (PDS)

Your PDS will be the most used component of Government 2.0. With your PDS you can:

- Manage which devices can access your PDS without your Federal ID Card
- Access your regular US Postal Service email
- Select and Manage who has access to your Medical Records.
- Access your Social Security Information[10]
- Access your IRS W2 and Earning History[5]
- Access your Selective Service History[5]
- Access other Agency Specific History as the agency data bases are transitioned to Government 2.0[5]
- Track and review which government agencies have accessed your information.
- Track and review which medical personnel have accessed your information.
- Review the security history of your PDS vendor compared to industry averages
- Select and manage which Government Authorized Agent (GAA) actually maintains you're PDS.
- Select and manage which GAA you use to calculate your income tax.

---

[10] These features will be enabled after the agency histories have been moved into your PDS as part of Government 2.0.

Depending on which PDS vendor you select you may also have additional for fee services, which are described by the vendor. These additional services may vary from vendor to vendor.

MANAGING ACCESS TO YOUR PDS

When you first registered at your FNO, you received two items, a Federal ID Card (FIC), and a onetime pass code. With your FIC you can access your PDS through any device which has a FIC reader. However, it is likely for most people that you will want to be able to access your PDS from your device. To do that you will need to use the one time pass code.

Using your Onetime Pass code

From any computer which is connected to the web you can use the One Time Pass code to set up that computer to access to your PDS. Entering the One Time Pass code, the FIS Service will show the PDS Management Page where you can create a username and password to access your PDS from that computer. When you have finished, the local PC will be updated to let you use your PDs. You can also alter the security levels you want to use for your PDS from that PC.

Once done, you can also use the management page to create additional One Time Pass codes. That way you can then set up your cell phone, or other devices to allow them to access your PDS as well.

Note that regardless of how many devices you link your PDS to, they all will have the same User Name and Password needed to access you're PDS.

LOST CARDS OR ACCESS TO YOUR PDS

If you lose your FIC, or can't remember your Username and Password, you can go to any FNO Office. They will help you get a new FIC card.

ACCESSING YOUR US POSTAL SERVICE EMAIL ACCOUNT

Your PDS is set up to send and receive email from an account under your name, according to your physical address.

As an example, let's use my physical address. I live at 3202 Woodrow Avenue in Richmond Virginia. Once I've registered to use my PDS my USPS Address Book Entry would show not only my address but also my USPS public email address. Let's say that address is Stephen_Imholt_1234.

That public email address is cross-referenced to my PDS. That way, if I move, my public Physical Address will change, but I don't have to change my USPS Email address.

The PDS vendor you picked will provide you with an email client like many available today. You can even change which email client you wish to use for your USPS account. And like most email clients today you can link your client to your other email addresses. You can even change your public email address to be one of your existing addresses.

How these things are done depends on which PDS vendor you choose. But all PDS vendors will have help functions online (and via web chat) if necessary.

Managing who has access to your Medical Record History

When you initially open your PDS, your Personal Medical Record History (PMRH) will be empty. You can begin the process of getting your PMRH loaded in several different ways. You can even decide not to load your PMRH at all. But be aware that some insurance companies are likely to discount rates for people who use them.

Today, when you change physicians there is a long process to get your medical records from your old physician. This process can take days and weeks to finish. Today's Electronic Medical Record is organized according to the physician, not you, so sometimes it's hard to get information for your own health quickly. One big change is how quickly your physician can begin to use your PMRH.

Here are the primary ways you can get your PMRH Loaded.

Using your PMRH folder management functions

Using the PMRH management functions you can select Add Physician which will walk you through a simple process. This process will allow you to find and select your physician and or group practice.

When you select your physician, an email from your PDS account is automatically generated and sent to the physician or group practice. The email informs them that you have identified them as being your physician. It also tells the physician's office that you are using the PMRH. Finally, the email also gives them an identifying code. The Physician's office can use that code to access your PDS. From there, the office can load your Medical Records to your PDS. The code which they can retain, will allow them to retrieve and view your medical records, and even to update those which are in process.

> ## Using your Federal ID card from your Physicians or Hospital Office
>
> Your physician will have access to a corresponding selection function. When you give his office your FIC card to verify, they can confirm that you have authorized him to update your PMRH.
>
> If you are taken to an emergency room, the ER staff can access your PMRH quickly using the same kind of process.

Accessing your Social Security Information[5]

When you get your PDS the folder flagged Social Security will be displayed as a folder under the Federal Government. But the folder probably won't open.

If you can't open the Social Security Information Folder, there is a good reason. It is because the Social Security Administration (SSA) isn't finished moving to a PDS based service. Your PDS vendor will enable this folder when the SSA is ready.

Until your social security information is set up, you can only do a few things.

> ## Request Social Security Contribution History

You can request your Social Security Contribution History. Doing this will cause the SSA to both access and load your contribution history to your

PDS. They also will tie a unique ID to the Social Security folder in your PDS. Each time the SSA accesses your information, this ID will confirm and control what and how the SSA can access and update your data in your PDS.

Load My Social Security Benefit History

This option is for individuals who are already receiving Social Security. This option will first load your social security payment history. It will also cause your earnings history to be loaded. Finally, the SSA will send you a summary of the benefits you have received from Social Security since you began receiving benefits.

Access to your IRS Information[5]

Once you have your initial PDS created, the folder called IRS will be displayed as a subfolder under the Federal Government Folder. Inside this folder are several folders. Like the Social Security folder, these folders will be enabled when the IRS information for people is ready to be distributed. These folders include

IRS History

This folder will contain subfolders for each year that the IRS had information needed to compute your earnings and taxes. Inside these folders will be not only the field copy of your 1040, but also the versions of all the supporting forms, as well as the W2 and 1099 statements from each organization which reported information to the IRS on your behalf.

Current Year IRS Information

This folder will contain the information for your current taxable year. This folder may be empty if you are retired or have not yet entered the workforce.

IRS Government Approved Agent (GAA) Information[11]

This folder contains information about your currently selected IRS GAA. Your IRS GAA is an approved organization who is authorized to process your tax return. The first time you open this folder, you will see the default vendor selected by the local FNO. You can change your selection to any other IRS GAA, prior to April 1. Once you've started to use an IRS GAA for the year, you must complete that year's tax returns using that vendor. More info is contained in the GAA help file for the vendor selected, or is available from the FNO.

ACCESS YOUR SELECTIVE SERVICE HISTORY[5]

If you are over the age of 18 and you have registered with Selective Service, your selective service history will be contained here. If you have not served, the information will consist only of your selective service application information.

---

[11] IRS GAA – An Internal Revenue Service Government Authorized Agent is one of the companies which have been approved to be IRS vendors. Each of these companies can assist you in completing the IRS reporting for the prior year tax filing. Not only do they provide the service to you, but these companies also provide confirmation to the IRS that your return is complete and accurate.

If you have not yet turned 18, you can use the document you will find in the folder to register with selective service.

ACCESSING OTHER AGENCY SPECIFIC HISTORIES[5]

Other folders will be added as more Federal Agencies move their information to your PDS. Each of these will go within the main government folder. Each time a folder is added, a notification will be sent to your USPS Email Account, letting you know.

The top folder in each agency folder will have a security document. This document tells you if the agency must request get your OK before accessing the folder. Most agencies will not need to request permission, but you can change the setting if you wish.

Either way each time a government agency accesses your folders you will be notified via an email notice.

Managing your Voting

One of the folders will actually contain a list of each election you voted in since your PDS folder was created. This is to both recognize your participation and to avoid potential voter fraud. This list DOES NOT CONTAIN which way you voted in any particular race.

The FNOs are operated based on local elections. For that reason, the current open election will be a hot link on the listing. There you can actually enter, save, and submit your vote on any number of local issues.

The current election will have a closing date/time. This is the date where you must have voted, saved and submitted your vote. This date could be used as a reminder notice as well.

Please refer to the Section on Managing the new FNO for further information.

Track and review what party accessed your information.

Your PDS can actually be accessed for a number of different legitimate reasons. But each time your information is accessed, or updated, a record is added to one of the logs in your PDS. This is to protect you. [12] Within the government wing is a set of folders which provide you with access to these logs for your PDS. These folders break the access down according to categories.

WHO CAN ACCESS OF YOUR PDS?

You

The first kind of access to your PDS is by you. Normally, the PDS folder will retain the last 4 times you accessed your PDS, but you can extend that

---

[12] If your PDS is currently the subject of a court ordered surveillance, your information may be accessed without your knowledge, UNTIL the surveillance is cancelled, or the order expires, at that point you WILL BE NOTIFIED that your information had been accessed, including all the information that you would have seen if it was accessed in the normal manner.

to retain more using the controls which show when you open the Access folder.

Your Physician, Hospital, and Insurance Providers

Each time a health insurance professional accesses your PDS, a log record is created. The record will identify who it was according to the approved id you gave them.

Insurance carriers will have access in the same way based on you giving permission.

The same controls that you can use to change how many logins of your information you yourself keep will also show you a list of providers to whom you have given access. When you change providers you can remove them from the list.

Government access of your information.

Each time a government agency accesses your information a log record is created. This record will identify the agency. The record also identifies the recorded purpose of the access. It will show which folder(s) were accessed, the date and time of the access, and whether the folder was updated.

In your PDS controls, you can set it up so that every time the log is updated you get an email.

Managing your PDS vendor

One of the primary folders in your PDS is the Management Folder. In this folder you can select which Government Authorized Agent/Personal Data Store (GAA/PDS) you use. The vendor you pick is the custodian of your PDS information. You will also be able to select which GAA/IRS vendor you want as your Income Tax Agent. And as other agencies transition to a GAA model, you can pick added Government Authorized Agents for additional government services.

When you first registered at your FNO you were given the option as to which PDS Vendor you wished to have. You can however, switch which PDS Vendor you have. You can do this through the Management Folder.

Selecting and Using your PDS Vendor.

When you first pick a PDS vendor you will be able to pick from among a number of different vendors. Each of these vendors may well offer other PDS services. These additional PDS services may include fraud protection, cloud storage space, credit card protection, etc. These services are fee based services, so if you select these, you will need to provide payment methods etc. The basic service for email, medical records are free, while other PDS Storage Services may not be free.

If for whatever reason you are unhappy with your PDS vendor, you may elect to transfer to another PDS vendor through these folders. You may find however that there is a limit to how often you can change a vendor without having to pay a fee.

MANAGING YOUR IRS VENDOR

It is assumed that there will be a re-engineered IRS. When that effort has reached the point where IRS GAA's become available, you will be able to pick one to be your GAA vendor. For most simple individual returns, it's expected that the service will be free, but if not, the prices will be clearly shown.

There is a significant difference between a company such as TurboTax as it exists today, and the IRS GAA vendor. Once you have successfully completed a GAA Tax return under Government 2.0, you are truly finished with your tax return for the year. Why?

Because part of the government certification process for GAA IRS vendors is a review of the actual vendor processing. Basically, the FIS will confirm that the IRS vendor is calculating the return accurately. This means that you will not be audited if you use one of these vendors. You will however, have the opportunity to do it yourself. But that will be subject to the IRS reviewing and perhaps auditing your return.

OTHER GAA VENDORS

As other agencies are re-engineered, there will be additional GAAs. This section has been left blank for that reason.

Management Information about your GAA vendors

Within your PDS folders are links to a few critical sets of information.

These include the obvious links for

Help

FNO Lookup

Links to your USPS Email information

Link to the Federal Information Service

Links to PDS Vendor Security Statistics (from FIS).

Here are brief summaries of each of these links.

HELP

There will be standard web based HELP functions. But, in addition to the basic help functions will be a link to a page which described what your FNO can do to help as well.

FNO INFORMATION LOOKUP

This page will contain information about your FNO such as operating hours, board members, services locally offered, remote services, and a link to the current FNO ballot, along with the election close out dates.

LINKS TO USPS EMAIL – AND HOW TO HELP MIGRATE TO MORE EMAIL

One of the links in your PDS will allow you to access the current USPS Email Web Page. There you can enter zip codes and find out if the residents are now already using Government 2.0 or not. In this way you can assist in migrating other users into making effective use of Gov 2.0.

LINKS TO THE FEDERAL INFORMATION SERVICE (FIS)

A separate link to the portal page of the FIS is available through your PDS. The Federal Information Service is the government agency in charge of implementing Gov 2.0. On this page you can link to the schedule for FNO office openings, current schedules for agency re-engineering projects, and links to forms for applying to be a GAA vendor.

LINK TO THE PDS VENDOR SECURITY STATISTICS

The FIS page will contain one very important link. This is especially true for those concerned about privacy.

In order to become an Authorized PDS vendor, each GAA/PDS vendor must meet the required criteria. This includes the requirement that each PDS vendor must produce submit statistics back to the FIS. The FIS will in turn publish these results through their portal on a monthly basis. This allows the consumer to determine whether the PDS vendor they selected is suitable for their security needs.

The initial set of stats that a PDS vendor must produce includes the following:

> # of PDS supported
> # of identified hack attempts on individual PDS
> # of unsuccessful penetration attempts on individual PDS
> # of identified successful penetration attempts on individual PDS
> % of PDS which are subject of court approved surveillance,

These statistics does not include whether your PC or smart phone was hacked, just the PDS service provided by the vendor.

MANAGING YOUR NEW FEDERAL NEIGHBORHOOD OFFICE

Your Federal Neighborhood Office is a new structure for our government. It cuts across local and state levels as well. It's easy to think of the FNO as a glorified FedEx/Kinko's or a more user friendly Postal Service. But you need to remember it's actually something very different.

When you first registered at the local FNO you were greeted by a service rep or concierge. They help you in being registered and getting your PDS set up. These people all work for a GAA vendor who is selected by the local FNO Management Board.

And you, along with the other citizens within your FNO service area, elect four of the five people who are that local management board. You can select them from people who live within your community, using the FNO Election Service.[13]

---

[13] The last member of the local management board is actually a representative of the GAAs who operate the local FNO. The local GAAs select a representative from the operating group to be a limited member of the board. This member only gets to vote in cases of ties between the elected four members

The FNO Election Service is specifically designed for the FNOs. Through it you can suggest items be added to the FNO Election ballot. These could include things such as changed operating hours. You will also be able to nominate candidates for the FNO Management Board. As it becomes available you can nominate candidates for the local of Federal Local Magistrate. You will also have the ability to support a recall petition for the board if needed.

STARTING OPERATIONS OF THE LOCAL FNO

For the first sixth months of operation the local FNO is managed by a board. This board is appointed by the following groups

> 1 member appointed by the US Congressional Representative
> 1 member appointed by each US Senator
> 1 member appointed by the Governor of the state
> 1 GAA representative appointed by the FIS.

These five members will comprise the local FNO board until the first election is held. That first election will be six months after opening. The first election will be for a term from the date of that election until the first Tuesday of the odd numbered year, when elections for the FNO board will again occur. The magistrate for the local Federal Magistrate court will be appointed by this board until the first local Federal Magistrate election is held. That will also six months after the board first begins operations.

The Federal Election Service

Every other year on the first Tuesday in November of odd numbered years, voters within the FNO can vote for how their local operations are to be delivered.

This includes

The members of the FNO board.

Whether the current FNO GAA should be extended.

Whether the current mail GAA should be extended.

Whether any other local GAAs should be extended.

Any ballot items added by the residents.

And periodically the local magistrate to sit on the Federal Local Magistrate Board.

At still other times (no more often that quarterly), the citizens of the FNO can vote on questions regarding operations of the FNO, including a recall petition of the local board and magistrate.

Voting on these issues is electronic through the individuals PDS.[14] Unlike regular election voting however, your vote is not "locked" in, until you lock it, or the deadline has passed. So you could enter a tentative vote, and then change it, up to the time you actually "lock" your vote. If you have entered a tentative vote, and the deadline for voting is reached, your tentative vote is "locked" at that point and counted.

---

[14] For individuals who have religious prohibitions regarding the use of technology, service reps at the local FNO can assist in voting to the degree possible, by actually recording the voters' preferences.

USING YOUR NEW FEDERAL NEIGHBORHOOD MAGISTRATE COURT

The Federal Neighborhood Magistrate Court (FNMC) is the second of the added parts of government with Gov 2.0. It extends the federal court system to have a local presence. In the FNMC the judges are called magistrates, and the position should be a part time role, not a career. Unlike the current federal court system, the judges won't be approved by Congress. Instead the magistrates will be approved by the FNO residents.

The FNMC is a local court which will rapidly take over as the initial court for what is today called administrative law.

Administrative Law is the part of US law that deals with government agencies. It handles the initial processes for agencies such as Social Security, the IRS, and the Veteran's Administration. There are many more agencies that have this type of law as well. As these agencies are re-engineered and decentralized, the functions of the administrative courts will be moved to the FNMC.

One specific type of case which can be immediately be placed within the Federal Neighborhood Magistrate Court are initial complaints regarding your civil rights including discrimination. This change is a result of laws which make infringement of civil rights a criminal rather than a civil matter. Lawsuits will be brought by the individual and/or local police to the local magistrate as a charge against an individual.

In the event you think you have a case of discrimination, you will be able to submit a complaint. The complaint will be initially heard by the local Federal Magistrate. At that point, he will walk you through the procedures, processes, and whether you may have reasonable cause to pursue the claim.

OTHER EVERY DAY FUNCTIONS IN GOVERNMENT 2.0

Your FNO office will be able to help you with many of the changes that are occurring with the deployment of Gov 2.0. Because the changes will be so drastic, it is very likely that you will have specific questions and concerns. It may deal any number of things about the federal and state governments do today.

For that reason, the Federal Information Service has a web portal. The portal allows you to request notifications about specific actions that are likely to occur within the future. Those notices will be sent to your email.

 The FIS portal also has an anonymous suggestion box for you to suggest changes to federal agency operations. This will allow the citizens to suggest ways to improve the effectiveness of federal agencies, including those which have already been re-engineered or have yet to be re-engineered.

The FIS portal also has a news component describing new FNO offices which are opening, upcoming and in process changes for the agencies currently being re-engineered, especially those agencies which affect the individual..

Finally the FIS portal has a whistleblower box to alert the FIS to abuses within the new Federal Neighborhood Offices, the new Federal Local Magistrate Court, and existing federal agencies.

HELP FUNCTIONS IN GOV 2.0

There are three major sources of help in Gov 2.0.

From your PDS vendor

Your PDS Vendor has a help function inside your PDS. It will help you answer many questions. Many PDS vendors also provide online telephone support.

From the FIS portal

For general questions regarding the Gov2.0 rollout, the FIS portal has a host of information. It will identify agency upgrades and consolidations. The portal will also assist with features being released to the FNOs and to PDS vendors

From your FNO

If you are unable to find what you need either through your PDS vendor or through the FIS portal, you can still contact your local service rep at your FNO. They will be able to help you, or at a minimum point you in the right direction.

RECEIVING FUTURE RELEASES

As new functions become available you will receive an email. It will have complete instructions about how to use the functions. For large enhancements, your FNO will have both classes and help to tell you about them. This includes new functions to your FNO as well as those which are added to all FNOs.

As new sets of data for you are added to your PDS you will be notified of that as well. And you will be kept up to date if you have requested information regarding specific updates and upgrades. You can request this through the FIS portal.

Generally, it is expected that there will be at least five years past the opening of the last FNO before the primary re-engineering of agencies is finished.

Right now there is no estimated date by which the upgrades for organizations[15] will be completed. There is also no timeframe for when the remainder of the federal government will be finished. By remainder is meant the intelligence agencies, defense department, state departments.

---

[15] Organizations includes companies, partnerships, corporations, and charities, religious and civic groups.

copyright Steve Imholt

CLOSING THOUGHTS

I hope this User Guide gave you something to think about how the Federal Government could work better. By writing as though it was all working, I hope I helped make it clear how badly organized the Federal Government really is. Today it operates as many independent siloed enterprises. The worst part is none of these agencies puts the citizen first.

The only place where it operates as a single unit it at the top of the pyramid. Even within the president however, isn't it obvious that sometimes that top person has no idea what is going on. This would be humorous, if it wasn't so wasteful.

If you are like most people, government has grown tiring. Worse, people have given up. People have tired of pundits and candidates yelling about how the other group is wrong, but never really telling you what they can actually do to address the issues of the day.

At a minimum this guide and the other books in the series do show that there is another way. You may not like it, and that's ok. But if you do, please help in changing this dream to a reality.

I am idealistic enough to believe that change is possible, yet realistic enough to know that without substantial participation from others, all my efforts are simply so much smoke in the ether.

The other two books, Volume 1 Vision and Volume 2 – Design and Implementation Planning like this Guide, both contain a single gargantuan assumption, that the vision has been sold. That is obviously not the case.

The Sale of the vision must precede the implementation.

Fortunately, the effort to go from vision to sale only (tongue in cheek) requires a few things.

- o   People who believe that this kind of change is needed
- o   People who believe that this kind of change is possible
- o   A venue for them to come together
- o   A means to educate the public on what can be accomplished.
- o   Individuals who are willing to step up and say "Now is the time", not later.

So here are a few ways that I hope may allow some of these ideas to move forward.

I'm running for the Virginia House of Delegates to see if this concept really has any merit. Look around the web for me. I'll start listing other people as they buy into the concept.

I'm continuing to write about what could be a new wave of actual representative democracy, However, that term must have a small "r" and a small "d", if it is to have any chance of success.

If I start to have some success in what is a true walk through the desert, we can begin to move forward with more things like a Foundation For Effective Government.

In the near future there will be a Facebook page for Government 2.0 where people who believe in the concept can go to contribute and refine.

Eventually something like www.gov2pt0.us will be a web site specifically designed for sharing approaches and details about the second editions of the various Government 2.0 books.

This website will be expected to frequently contain the work products of the foundation, and will end up being owned and operated by the foundation staff.

Until these web sites are enabled, suggestions for sharing approaches and details along with other seminal activities will be contained as a part of the SteveImholt.com website. I suggest you look it up.

I want to thank everyone for taking the time to read and evaluate the contents and approaches recommended through these books.

GLOSSARY

GOVERNMENT AUTHORIZED AGENT (GAA)

A Government Authorized Agent (GAA) is a general term used to describe one of two kinds of people. One type is a federal employee. The other type is an employee of a GAA vendor. A GAA vendor is a company which provides a specific service to the Federal Neighborhood Office.

These services are delivered according to predefined standards and procedures. Companies like TurboTax in the future will be examples of a GAA vendor.

Another type of GAA vendor are the PDS vendors. Companies like Amazon will provide these services. Even the janitorial staff could be GAAs. They may also include for specific FNOs things like job training, employment counseling etc.

PERSONAL DATA STORE (PDS)

At its core the PDS is a cloud storage account. Initially it will be available from a variety of the larger commercial vendors such as HP, Dell, IBM, etc. It is important to remember the PDS itself is owned by each person.

Each PDS contains as much or as little added optional storage and data as the owner can afford. The owner can include links to personal social networks such as Facebook. The PDS will always have folders for the different kinds of information which the US government uses. Beyond that the individual owner decides how much or little is used.

Because the person owns the PDS (kind of like a car), each person can decide where to "park" their PDS. The PDS owner can select to use the PDS storage package of any GAA. The GAA standards for the PDS include standards for both security and ease of access. The security standard protects the individual from unlawful access by hackers. The standards also protect against unlawful access by a GAA.

Before all the PDSs are in place over 100 private companies will have their own version of PDS accounts. This will result in 325 million cloud storage accounts, or on average less than 3.25 million people per company.

Some people cannot afford to pay for a PDS with added functions. Others will decide not to own their own PDS. For these groups of people the PDS will contains only the Government, email, and medical Insurance data.

Individuals at any time may decide to begin using their PDS. At that time the individual areas can be added by the vendor they choose or which already has the government version. At first, people will be will be able to initially access their PDS using their Federal Identification Card (FIC) at any Federal Neighborhood office (FNO).

FEDERAL IDENTIFICATION CARD (FIC)

The US Federal Identification Card (FIC) is an ID card issued to every person in the US. The FIC uniquely identifies the person. It will have state of the art encryption on its contents. People can use the card to get into their PDS. The card can contain multiple credit accounts reflecting amounts credited to the individual from various US Federal Agencies or even personal bank accounts of the individual. It may be used for example as a SNAP card.

In all about 325 million FICs will be issued to people in the United States. The administration of the physical cards will be managed through the Federal Neighborhood Offices (FNO).

The cards themselves will be manufactured by a number of companies. Each company is a GAA who must meet the standards defined by the FIS.

FEDERAL LOCAL MAGISTRATE COURT (FLMC)

The FLMC is a new level of the Federal Court system. It is the place where two kinds of court cases begin. It is the place where the initial adjudication of federal misdemeanors. Most of these cases are in what is today called administrative law. Today, most of these cases actually begin in the executive branch. THE FLMC is also the place where cases regarding privacy and discrimination crimes begin. The FLMC is expected to operate at the community level. It is expected that it will be only a part time court. In some places it may even be a virtual court. The magistrates who hear the cases are members of the local community who have had specific training.

copyright Steve Imholt

FEDERAL INFORMATION SERVICE (FIS)

The FIS is a new federal department. It has ties to both the Government Accountability Office (GAO) and the Office of Management and Budget (OMB)[16]. The FIS is responsible for making sure the FNO operations procedures are complete and work well. The FIS also sets the standards for the new features of the Federal Government. This includes the standards for PDS storage vendors and the FIC. The FIS also sets up the standard service contracts for vendors providing services to the FNO and the FNMC. Finally, the FIS manages the projects and programs which will deliver Gov 2.

---

[16] While it is critical that the FIS be at a level to command both the support and the input from the other federal agencies, it is just as important that it have a link to the legislative branch, Whether it is an additional legislative branch organization, or a part of the executive branch it must have a presence in both components of the Federal Government. So it could just as easily be part of the OMB and dotted line to the GAO.

ABOUT THE AUTHOR

Steve Imholt is a long time project manager, IT architect, designer and developer. Steve has been both witness and part of the evolution of the Information Age. He has worked with computers since the days of punches cards and designed services to operate within the cloud.

Steve's life has been essentially a geek version of Forrest Gump. Steve has had a front line view of the evolution of technology.

In the 1970's Steve gained his initial experiences in healthcare as a systems engineer trainee at EDS. This was the time where the Medicare processing expansion hit its stride. This was followed by a short time where he worked at Merchant's National Bank. It was here where he helped implement the banks first interest bearing checking accounts. This was followed by a period where Steve was the DP manager for a grocery store chain just as Universal Product Codes were being introduced to the grocery industry.

 In the 1980's Steve gained even more experience in the health care field in both hospitals and insurers. This was a time where there were rapid changes in health insurance. New concepts such as HMOs and PPOs were added by insurance companies. New frameworks such as Certificates of Need and health networks were added. These were all failed attempts to get health care costs under control.

It was during this time that Steve dabbled in local politics being elected to a local school board. There he served as co-chair of the finance committee. After moving because of job changes, Steve then served on a community library board. Steve developed a better understanding of political groups because of these experiences. He also saw how politics was very different between the levels of government. Steve saw how special interest groups of all types were beginning to affect all the layers of government.

Technology in the 1980s and 90s was the age of the Client Server world. Along with companies like HP and Compac, Steve's skills changed to be a force within that world. In that world, he learned about the real complexities of safely disposing of hazardous materials. He also learned about the pork barrel regulations. Regulations such as excluding sites where Congress received large contributions. It was these kinds of regulations that blocked moves to clean up some parts of the environment. That was the Environmental Protection Agency of the time. Sadly, it is also the EPA of today, simply with a different set of special interests.

In the mid-90s allowed Steve was asked to manage the development of a mortgage loan origination system. These loans were for manufactured housing, which most people call mobile homes. The origination system was a way for a REIT[17] to create Securities. Because mobile home loans usually were less than high quality, they became part of the Subprime

---

[17] REIT stands for Real Estate Investment Trust. These were companies specifically set up to create Mortgage Backed Securities, which fueled the investment expansion of the 1990s and early 2000s.

Mortgage Loan market. It was these types of Securities many people blame as a major cause of the Great Recession.

From there after failing at starting his own IT business, Steve became a software Architect. He took that role beginning in the summer of 2001 working for Tauck which is the world's oldest group travel company. They were located outside of New York City in Westport Connecticut. Like many others Tauck was devastated by 9/11. In the course of two weeks, that company like others saw the size of its business drop in half. Steve was fortunate enough to be able to continue working with them until 2003. At that time, the company could no longer support the size and scope of their re-engineering efforts. As a result, they offered and Steve took a retirement package.

Finally in the early 2004's Steve began work with Hewlett Packard on the New York City ECTP program. ECTP was a NYC program to update the city's emergency response systems. There Steve learned about the bid and development processes used by government agencies. Those same processes were seen during his time at the United States Postal Service Project. As Steve helped an old friend bid on a Veterans administration project he learned that these processes were not unique to HP. Instead the processes for both bids and development are dictated by the government rules.

Multiple HP projects followed. While a few dealt with the commercial sectors, most projects continued to be in government. Some were in national defense or hospitals. Other projects were in state government and education. As Steve's experiences proved increasingly valuable, Steve provided trouble-shooting support for several HP projects. These projects were generally those which were part of acquired companies. After the acquisition was completed was when the troubles became obvious. In most of these situations, at least part of the source of the trouble was in the flawed bid and project award processes themselves. In others it was the demand that changes have little or no effect on existing processes. The common thread was that all of these projects were inconsistent in the requirements and the solutions, in large part because of the bid and development processes.

All of these projects led him to the conclusion that the development process was no different between state and federal government. This was unfortunate, because these processes do not work.

Steve as an architect began looking for the real core issues confronting integrating technology with government. Why government systems are so difficult to successfully change or replace? Can the public in concert with business deliver a better way for technology to assist government?

All of these experiences directly lead to this series of books.

Steve lives in Richmond Virginia with his wife Toni. Together, they raised three girls and a boy. Their children are now adults each with successful careers. These careers include one child who is in IT, another in medicine, a third in education and the last is an attorney. Basically because of mostly her efforts the Imholt's have hit today's version of a child-rearing grand slam. Now Steve makes woefully inadequate attempts to help mentor three of the four granddaughters, while his wife succeeds in rearing the next generation as the preeminent grandma at our home now renamed Gram's House.

www.ingramcontent.com/pod-product-compliance
Lightning Source LLC
Chambersburg PA
CBHW070947210326
41520CB00021B/7091